YOUR KNOWLEDGE HAS VALUE

- We will publish your bachelor's and
  master's thesis, essays and papers

- Your own eBook and book -
  sold worldwide in all relevant shops

- Earn money with each sale

Upload your text at www.GRIN.com
and publish for free

Sören Noack

Environmental Impact of Jeans Laundries in Northeast Brazil

GRIN Verlag

Imprint:

Copyright © 2008 GRIN Verlag GmbH
Druck und Bindung: Books on Demand GmbH, Norderstedt Germany
ISBN: 978-3-640-86615-1

This book at GRIN:

http://www.grin.com/en/e-book/167981/environmental-impact-of-jeans-laundries-in-northeast-brazil

ENVIRONMENTAL IMPACT OF JEANS LAUNDRIES IN NORTHEAST BRAZIL

Sören Noack

Study Project within the Master program in Environmental and Resource Management at the Brandenburg University of Technology Cottbus

Summer Term 2008

Cottbus, September 9, 2008

Preface

These days terms like *climate change, environmental protection* and *sustainable development* have become buzzwords widely mentioned in the media. But instead of hoping that scientists will come up with feasible solution to solve today's most important environmental problems we are facing, it might be better to think about our own consumption behavior. Before blaming the so-called *developing countries* for their environmental destruction, we better ask how our daily food is related to the destruction of the Amazon rainforest. How many liters of water does it take to produce the pair of jeans we are wearing right now? Not to mention about working conditions in the textile factories.

This paper focuses on the environmental impacts of the jeans industry in Northeast Brazil, based on a three-months project stay in the region. My experiences during this time encouraged me to think about the relation between consumption behavior in the so-called *developed nations* and environmental damages in *industrializing countries*, even though this particular case is different in so far, that due to the disparities within Brazil, consumers of more developed regions cause problems in less developed regions within the same country.

At this point I would like to use the opportunity to thank all persons that helped me during the last months in various ways. Without the support by the ASA-Program and BFZ as project sponsor, this study project could not have taken place. Further, I like to thank Dr. Becker for this efforts of being my supervisor at Brandenburg University of Technology. I also thank the experts from ITEP and my colleague Florian Heiser for their valuable contributions, as well as the involved laundries for their willingness to cooperate.

Contents

1 Introduction   **1**

2 Background   **3**
    2.1   Development in Brazil 3
    2.2   Development of Northeastern Brazil 6
    2.3   Importance of the textile industry for local economy 6
    2.4   Environmental issues 8

3 Environmental Problems   **11**
    3.1   Water Pollution 11
       3.1.1   Applied treatment technology 11
       3.1.2   Characterization of the wastewater 13
       3.1.3   Remaining problems 14
    3.2   Solid Waste 14
    3.3   Steam Production and Energy Efficiency 15
       3.3.1   Fuel Choice for Steam Production 16
       3.3.2   Energy flows and losses in laundries 18

4 Future of Jeans Laundering in Caruaru and Toritama   **23**

5 Summary and Conclusion   **27**

List of Figures

2.1 Human Development in Brazil   5
2.2 Human Development in Pernambuco   5
2.3 Typical scenery of small laundry in the backyard   9

3.1 Treatment technology 12
3.2 Treatment in two different laundries 12
3.3 Solid waste treatment 15
3.4 Fuel used in laundries 16
3.5 Fuel costs 17
3.6 Boiler systems to generate hot water and steam 18
3.7 Distribution of water temperatures needed in a selected laundry 20

4.1 Impact on the local population 25

List of Tables

3.1 Wastewater characteristics for a selected laundry 13

Acronyms and abbreviations

ASA	Work and Study Exchange ("Arbeits- und Studienaufenthalte")
BFZ	Training and Development Centers of the Bavarian Employers' Associations
BOD	Biological Oxygen Demand
CDM	Clean Development Mechanism
COD	Chemical Oxygen Demand
CPRH	Agency for Environment and Water Resources of Pernambuco
FDI	Foreign direct investments
FIEPE	Association of Industries in Pernambuco
HDI	Human Development Index
ITEP	Technical Institute of Pernambuco
NGOs	Non-governmental organizations
SINDIVEST	Business Association of the Textile Industry
UFPE	Federal University of Pernambuco
UN	United Nations
UNDP	United Nations Development Programme
UNEP	United Nations Environmental Programme

1 Introduction

Recently, environmental issues receive more and more attention by the public worldwide. Nevertheless, people often expect politicians and industries to come up with solutions, especially in the context of climate change mitigation. The *inconvenient truth* however is that there is a clear link between consumption behavior and environmental degradation. Even worse, in many cases consumers in developed nations are responsible for impacts affecting people in less developed regions. One example is clothing. Nearly everybody has got at least one pair of jeans, even though most people have no idea about jeans production and the impact thereof.

The aim of this paper is to present environmental impacts of the jeans industry in the case of *jeans laundries* in Northeast Brazil and to discuss possible solutions under the given socio-economic constraints of this region. First, some general background information is provided in order to point out the outstanding importance of the clothing industry for the local economy. This importance explains why for several years an unwritten agreement between local politicians and companies in form of a *devil's deal* existed, that freed enterprises not only from environmental and labor regulations, but also tax payments.

Interesting is that in this particular case, it were not the authorities, but an entrepreneur that started looking for possibilities to reduce the environmental impacts. This had been the trigger for authorities to become more active again. Chapter 3 contains an analysis of the environmental problems caused by wastewater and residuals, as well as applied solutions. A special role play issues of energy efficiency in steam and hot water production, based on a detailed analysis of one of the biggest laundries. Finally, an outline for the future of jeans industry is provided.

2 Background

This chapter provides an brief overview on the existing disparities and its implications on environmental awareness within Brazil in the first part. A more detailed view on the situation in the federal state of Pernambuco and the development of the area of the textile industry cluster of Caruaru, Santa Cruz do Capibaribe and Toritama is presented in the second part. Having this background knowledge is helpful for the understanding of issues described in the following chapters.

2.1 Development in Brazil

The Federal Republic of Brazil is the fifth biggest country of the world and the largest in South America. Compared to other developing countries, indicators of living standards are relatively good, though the country still faces problems of the developing world, for example a relatively fast growing population and especially a rapid urbanization. Within the last 40 years, the share of urban population increased from 40% to 80% (Ramakrishna et al., 2003), causing social problems in urban agglomerations. Brazil is still one of the most unequal countries, especially in regard to income distribution and land possession, which are highly concentrated. The richest 20% of the population accumulate 66.1% of the total income, while the poorest 20% of the population receive only 2.3%. Less than 3% of all farming land is available for small farms (< 10 ha), but large farms (> 10,000 ha) occupy more than 40% of the farming land (Seroa da Motta, 2002).

In the following, the Human Development Index (HDI) is applied to visualize regional differences in development within Brazil. The HDI takes three dimensions into consideration: (1) life expectancy, (2) knowledge (average adult literacy and years of schooling), and (3) the standard of living (real per capita income). These three indicators influence the total HDI by one third each, so that the final index may sum up to the value of 1.0 in the optimal case. The United Nations Development Programme groups countries according the Human Development Index into three categories (UNDP, 2007):

- Countries with high human development (HDI $\geq$ 0.8)
- Countries with medium human development (0.5 $\geq$ HDI $\leq$ 0.8)
- Countries with low human development (HDI $\leq$ 0.5)

Concerning Brazil, the country's average HDI value increased from 0.789 in the year 2000 to 0.800 in 2005 (UNDP, 2007), i.e. its classification changed for the better from a country with medium to a country of high human development. However, considerable regional disparities can be observed (Figure 2.1 on the facing page). In general, one can state that in South and Southeast the living standard is the highest. The Southeast, especially the conurbation of São Paulo, is Brazil's economic and industrial center. All three Southern states (Paraná, Santa Catarina and Rio Grande do Sul) are characterized by a strong influence of European immigration in the 19th and the beginning of the 20th century. On the other hand, the Northeast is often considered as *poorhouse* of Brazil and faces problems due to structures established with the colonization since the beginning of the 16th century. Except from tourism at the coastline, on the countryside sugar cane plantations and processing industries still play a vital role as one of the few possibilities for the region to provide employment. Due to low development in connection with hot and dry climate, economic alternatives are hard to find.

Bearing this in mind, the importance to strengthen environmental awareness and education becomes evident, since the acceptance of lowering the environmental quality in turn for economic growth increases especially among the lower income-classes. Many companies are lacking of knowledge concerning main initiatives on global environmental issues, though a great majority already recognizes the relevance to their business activities. Even the current political agenda concentrates rather on economic growth and reduction of social gaps. Environmental concerns have to be balanced against development and equity issues (Seroa da Motta, 2002). Consequently, authorities may refrain from pursuing environmental policies, as pointed out in the following.

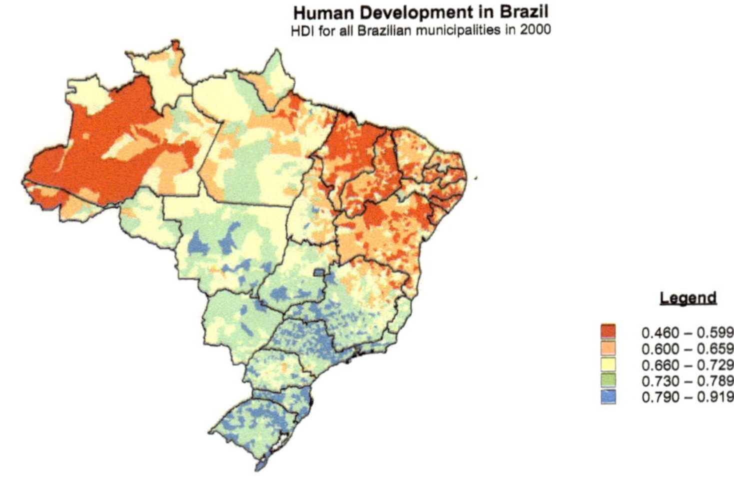

Figure 2.1: Human Development in Brazil
(Source: UNDP, 2003)

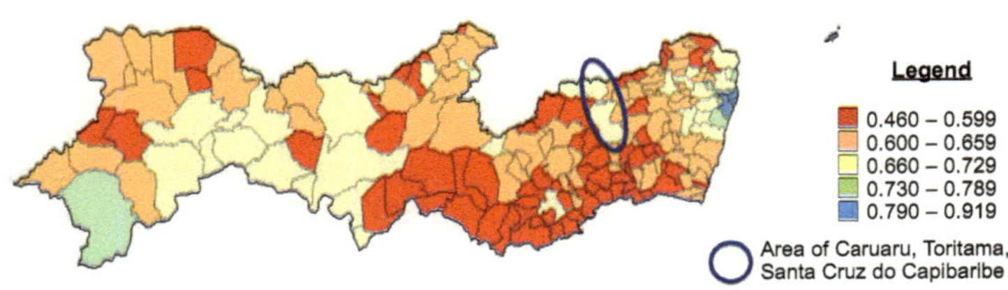

Figure 2.2: Human Development in Pernambuco
(Source: UNDP, 2003)

2.2 Development of Northeastern Brazil

Pernambuco is one of the nine federal states in Northeast Brazil and belongs to the least developed parts of the country. Besides the touristic areas along the coastline, in people's perception the countryside is associated with poverty and droughts. Regarding the poverty level, the region can be compared with Central African countries. Approximately 75% of the inhabitants, and respectively 60% of the urban population, are living at or even below the poverty line of 1 US$ per day (Meier & Wahl, 2005).

As shown in figure 2.2 on the previous page, most municipalities in Pernambuco are either colored dark or light red, i.e. have a rather low HDI between 0.460 and 0.659. Caruaru, Santa Cruz do Capibaribe and Toritama are classified in the next higher category (beige) with HDIs from 0.660 to 0.729 and therefore higher developed than their surroundings. Except for textile industry, Caruaru as second most important city of Pernambuco serves as an important trade center for the region. Furthermore, climate conditions still allow agriculture. Higher developed regions can be only found in and around Recife (capital of Pernambuco) at the coast, and Petrolina in the Southwest. Petrolina gained its boom with the construction of the Sobradinho dam, which not only provides electricity, but also allows irrigation and made the area to become a wine producing region.

2.3 Importance of the textile industry for local economy

Providing around 140.000 jobs, the garment industry is the second most important economic sector. The major cluster of clothing industry emerged in the municipalities of Caruaru (289,086 inhabitants), Toritama (29,907 inhabitants) and Santa Cruz do Capibaribe (73,667 inhabitants). Located ca. 120 km from the capital, Recife, this area comprises a population of almost 400,000 inhabitants (IBGE, 2007) and became Brazil's major jeans manufacturing center, supplying ca. 18% of the country's jeans production. This industrial sector is furthermore characterized by a high fraction of small and medium sized family enterprises. In several cases, one member of a family owns the dressmaking company, while another family member takes responsibility for the laundering activities. Ongoing market liberalization puts pressure on the local textile industry due to cheap imports of mass products from Asian countries like China.

During the first half of the 1990s, almost all of the firms in the sector used to be informal, i.e. operating without proper registration and paying taxes. To the mind of the owners, this resulted from bureaucratic barriers and high tax burdens (Meier & Wahl, 2005). But for years also the government of Pernambuco avoided to enforce labor, environmental and tax regulations. Officials were simply afraid to disrupt the local economy. Due to the absence of tax payments however, the government felt not obliged to invest in improving the infrastructure. The result was a *devil's deal*[1] between government officials and entrepreneurs (Almeida, 2005). Another outcome of informality was the inability to access bank credits and support from governmental agencies for vocational training, as well as joining business associations. This situation started to change with the election of a new board of the business association Sindivest[2] in 1997, that started to take care of the interests of local businesses and successfully lobbied for tax reductions from 17% to 4% on sales. Officials came to the conclusion that "*a larger number of businesses which pay lower taxes might net the country more than a small number of businesses paying high taxes*" (Meier & Wahl, 2005). Regularization also allowed firms obtaining support in order to improve product quality and diversity. Until the middle of 1990s, clothes from the region were known to be rather cheap and of low quality. Only a few companies were able to sell their products on other important markets in Brazil. This situation improved already, but needs urgently further efforts to defend its market share with ongoing liberalization of the Brazilian market and also conquer other markets to ensure that this sector can survive.

---

[1] Tendler (2002) uses the term *devil's deal* to describe the deal between stakeholders and local entrepreneurs in the form of "Vote for me, so I won't do anything to make you comply with labor, environmental and tax regulations". Involved politicians often assume that in such a way they help small businesses to survive.

[2] Sindicato das Indústrias do Vestibular, Business Association of the Textile Industry

2.4 Environmental issues

Compared to the production of other clothes, denim production has a strong impact on the environment. Over the whole production cycle from planting cotton to the end product, Chapagain et al. (2006) estimated the average of 10,850 liters of water consumed per pair of jeans in total. Most of the amount of needed water is related to cotton growing. For jeans production in Caruaru and Toritama, the main impact is caused by the washing process. In this last step of production, detergents, conditioners and other chemical substances are applied to obtain jeans with the desired look. The exact amount of water needed varies on the individual washing process, mainly influenced by the different design of jeans. Values on water consumption are ranging between 60 and 100 liters per pair of jeans (Almeida, 2005). Data provided by one laundry yields to an average of 70 liters per piece. Given a total output of one million jeans every month by all companies, implies not only the consumption of up to 60-100 million liter of water, but also means that the same amount of wastewater needs to be treated in order to avoid environmental damages and risk to health. Furthermore, water and wastewater treatment are also considerable cost factors.

One of the first steps towards environmental improvements was taken more or less by accident after the droughts in 1999, when the price for water increased dramatically and laundries' owners had to organize transport of water by trucks from nearby cities. Interested in an economical solution to reduce the high costs for water, the owner of one of the largest laundries in Toritama, contacted SINDIVEST. Finally, SINDIVEST established the contact to BFZ[3], that was interested in developing a technology to recycle water and control water pollution. BFZ came up with a low-cost solution which was 70% less expensive than already existing ones, since it used only material available in the region.

The other important factor was the nomination of a new public attorney in Toritama in 2001. Influenced by education that promoted *"the importance of environmental law to achieve sustainable development"*, he *"was shocked by the lack of compliance of local firms with the labor regulation and the environmental law"* (Almeida, 2005). At the same time, CPRH, the state agency for environment and water resources, started their investiga-

---

[3] Berufliche Fortbildungszentren der Bayrischen Wirtschaft, Training and Development Centers of the Bavarian Employers' Associations

tions on environmental pollution by jeans laundries. Since the low-cost solution of BFZ was available, the laundries were obliged to implement a wastewater treatment system. Most of the treatment facilities have been planned and adopted by ITEP, the technological institute of Pernambuco. At present, all laundries possess their own wastewater treatment system. As side effects, also the handling of chemical substances and general waste treatment improved. Furthermore, workers benefit from increased security, even though there is still need for better protection. Nevertheless, all efforts undertaken so far have to be considered as first step and require further improvements. A more detailed analysis of the environmental aspects, including issues of energy efficiency, are addressed in the following chapter.

Figure 2.3: Typical scenery of small laundry in the backyard
(Source: own)

3 Environmental Problems

3.1 Water Pollution

Before the installation of treatment systems, the blue colored and contaminated wastewater was directly discharged to the local channels, creeks and rivers. Everything ended up in the most important river of the region, Capibaribe, which feeds a reservoir supplying many cities in the region. According to Almeida (2005) 83 towns, including Caruaru as most populated city, had been affected. This issue showed the urgent need to apply developed low-cost technology.

3.1.1 Applied treatment technology

The system consists of a minimum of three different tanks for the purification of the water for equalization, flocculation and filtration processes. First, all water from laundering is collected in the equalization tank (Figure 3.1a). In the second step, water enters into the flocculation tank. During the treatment, a flocculant such as aluminum sulfate is added and the mixture is agitated by a small engine. As result of precipitation, aggregated particles settle down and the purified water remains in the upper half of the basin. As third and last step, purified water passes a sand filter (Fig. 3.1b) and can be either used again for laundering or is discharged. The remaining sludge is released later to dry in the sun. Often, the same basin can be used after the cleaned water had passed the sand filter. Sand and sludge are then removed, and the sand filter is getting prepared for the following treatment.

Main advantages of the system are low costs and its flexibility to adopt to the needs of each company. Depending on the available space, tanks and filters can be arranged

Figure 3.1: Treatment technology: Wastewater flowing into the equalization tank (a) and Sand filter (b)

(Source: own)

Figure 3.2: Treatment in two different laundries
(Source: own)

either horizontally or vertically. In relation to the size, one or more tanks have to be added (Figure 3.2 on the facing page). The whole system can be constructed with cheap local materials such as bricks and sand, and workforce. Since labor costs in Northeast Brazil are quite low, overall costs remain on a reasonable level. On the other hand, it must be absolutely clear that such a technology is not as efficient as most advanced treatment systems applied in South Brazil or industrialized countries. But in fact there is no feasible alternative available at the moment. The applied low-cost systems have sufficient potential to improve the environmental situation and fit to current need of small and medium-sized enterprises.

3.1.2 Characterization of the wastewater

Laundering processes require the use of different chemical products such as sodium chlorate, sodium hydroxide, hydrogen peroxide and detergents. Consequently, effluent characteristics are subject to high variations, as presented in Fig. 3.1 for the Biological Oxygen Demand (BOD), Chemical Oxygen Demand (COD) and pH. Applied physical-chemical treatment may remove up to 70% of BOD and 71% of COD loads. Though these results are remarkable, the treated wastewater would still not meet the standards of developed countries. Another problem are increased concentrations of alumina, calcium and magnesium, as alumina sulphate and calcium hydroxide are added during the treatment, limiting the potential reuse of the water (Santos, 2006). However, at least in theory, the pre-treated wastewater could be discharged to the sewer network if there was an existing collection and treatment station for domestic sewage.

Table 3.1: Wastewater characteristics for a selected laundry

Indicator	Untreated Effluent			Treated Wastewater			Removal Efficiency
	Average	Min	Max	Average	Min	Max	Average
BOD (mg/l)	1032	26	4900	307	216	380	70,25%
COD (mg/l)	2885	129	10579	827	627	1100	71,33%
Chlorides (mg/l)	4746	920	12675	1572	1230	1910	66,88%
pH	8,2	5,7	10	6,8	4,3	9,2	

(Source: Santos, 2006)

3.1.3 Remaining problems

Currently, all jeans laundries are obliged to have a wastewater treatment system. However, as seen during the field studies, the existence of treatment system does not necessarily imply that the treatment is done adequately, resulting in removal efficiencies of organic loads in the range of 30-40% (Santos, 2006). One reason might be insufficient knowledge of the responsible staff, the main reason however is the lack of control. During the time of the field studies from September to November 2007, only researchers from UFPE were taking samples of 3-4 selected laundries every two weeks. These companies are involved in the time schedule of sample taking, besides the fact that they do not have to fear consequences in case of not complying with respective environmental regulations. According to locals, the responsible organization of the government, CPRH, checks only whether proper treatment systems have been installed. Due to this lack of monitoring, wastewater treatment is seen as a pure cost factor and not always done.

Another observation during the field studies was that with the installation of wastewater treatment systems the pollution problems have only been shifted from the contamination of water bodies to potential soil contamination, due to improper handling of the sludge resulting from treatment.

3.2 Solid Waste

While during the first phase the water related problems had been focused, also issues of solid waste treatment are receiving more and more attention. As Almeida (2005) pointed out, with the implementation of wastewater treatment systems, simple improvements regarding the storage of chemicals and disposal of packing material have been realized. Furthermore, with the program "Lavar sem sujar"[1], ITEP and several other associations tried to promote the separation of solid wastes in jeans laundries into six categories: wood, plastics, glass, metals, wood and residual materials (Fig. 3.3a). This has to be seen as prerequisite for the introduction of separate waste collection, an issue that the town hall is currently planning to implement.

---

[1] Lavar sem sujar = Wash without pollution

Figure 3.3: Solid waste treatment: Waste separation (a) and sludge disposal (b)
(Source: own)

Another problem is however the inadequate treatment of residuals deriving from the waste water treatment process. Due to nonexistence of an industrial landfill, the content of the sand filter had been removed and either illegal displaced somewhere in the landscape or put on bare ground until the rain washed out the chemicals (Fig. 3.3b) and then the "cleaned" sand was used again in the filter. This means nothing else then shifting the pollution problem from issues of wastewater to solid waste disposal. Finally, in November 2007, the town hall of Caruaru decided to allow the disposal of these residuals on their municipal landfill and during the next months, this issue should be solved.

3.3 Steam Production and Energy Efficiency

During the field studies, I concentrated on issues related to energy efficiency of steam production and distribution. Jeans laundries are using steam at three stages during the laundering. Usually, inside the washing machines hot steam with a temperature of about 200°C is mixed with unheated water (ca. 25-30°C), in order to reach the required temperature. The exact temperature depends on the specific washing process applied and is normally in the range between 40°C and 95°C. Furthermore, steam is required to dry the textiles after laundering in the driers and additionally ironing of jeans also requires steam for flat irons. During all stages, including the steam production, energy is unnecessarily lost. This is not only a financial issue, but also leads to significant environmental impacts.

3.3.1 Fuel Choice for Steam Production

The vast majority of companies uses wood in their boilers. Most of the owners reported to pay about 25 Brazilian Real (R$) or even less per cubic meter of firewood, which is frequently illegal cut. Since control mechanisms are lacking, illegal cut wood is the most attractive option. All other alternative fuels are more expensive. For example, licensed wood of better quality costs ca. 40 R$/m^3. Other options include natural gas, bagasse (Figure 3.4) and the use of solar energy to preheat the water before entering into the boiler. Especially bagasse could be a promising alternative, as it can be considered as residual derived from processing sugar cane to alcohol and sugar, another important sector in Northeast Brazil. In theory, one ton of bagasse could replace up to seven cubic meters of wood (Cabral, 2007).

Figure 3.4: Fuel used in laundries: Bagasse briquettes (a) as alternative to firewood (b)
(Source: own)

However, there exist already several other applications, for instance as animal feed, and currently research is done on the production of alcohol from bagasse. Therefore, the available amount would not meet demand of all laundries of the textile cluster. For one of the biggest companies, five several fuel option have been analyzed (Figure 3.5 on the facing page): Thick and fine wood, bagasse briquettes, babassu and petcoke. Petcoke is

Fuel type	Specific Costs	
	[R\$]	[€]*
Thick Wood**	0.0624	0.0240
Fine Wood	0.0770	0.0296
Bagasse	0.1706	0.0656
Babassu	0.2367	0.0910
Petcoke	0.1763	0.0678

* 1€ = 2,60 R\$
** including reuse of condensate

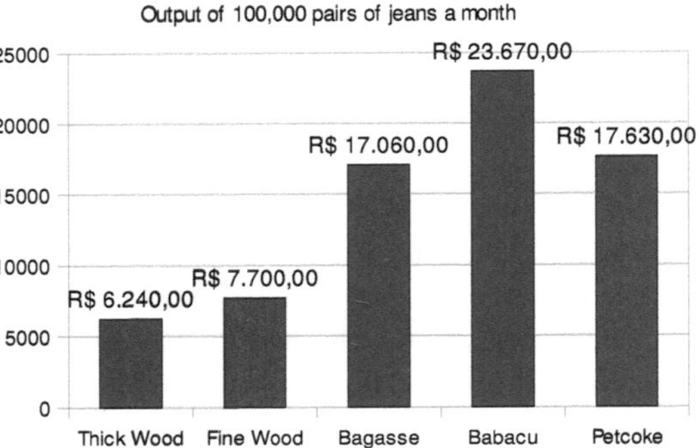

Figure 3.5: Fuel costs
(Source: own creation based on personal communication with a laundry)

a residual from the petroleum industry and represents the only viable fossil option, since natural gas by pipeline is still not available in the region. Babassu is a palm tree, the nuts can be used to produce oil, which is a promising renewable fuel source. Furthermore, wood is distinguished in small and thick wood. Important to note is that for the potential use of thick wood, it is assumed that the condensate from the boiler and driers is reused, so that the fuel consumption can be reduced.[2] Therefore, it is not surprising that in the comparison of the considered fuel types thick wood is the most competitive possibility. However, the analysis of specific costs per pair of jeans also clearly shows that bagasse (0.17 R\$ per pair of jeans), considered as most attractive alternative to firewood, is more than twice more expensive than the use fine wood (0.08 R\$ per pair of jeans). Even though a difference of 0.09-0.10 R\$ in marginal costs might be not that huge, given a monthly output of 100,000 jeans, it amounts in higher fuel costs between 9,360 R\$ (fine wood) to 10,820 R\$ (thick wood) compared to the use of wood.[3]

---

[2] Data provided by the company only considers the reuse of the condensate in the case of thick wood. In principal, the reuse of condensate for the other fuel options is possible too, but was not considered by the company.

[3] Since the company did not disclose its exact cost structure, the share of energy costs on total costs can not be determined. Nonetheless, these numbers provide an estimation of minimum increase in costs caused by fuel changes.

One of the few laundries using bagasse pellets is a medium-sized laundry and jeans designer in Caruaru. While observing their laundering process for one day, 241 pieces of jeans had been washed, causing the consumption of 240 kg of bagasse and 13.000 liters of water. This corresponds to marginal fuel costs of 0.21 R$ per pair of jeans, which is somewhat higher in comparison to the data in figure 3.5 on the previous page. As it had been only possible to visit this company once, it is difficult to say in how far this is a typical average value. However, it serves as example that the use of bagasse seems to be economic feasible.

3.3.2 Energy flows and losses in laundries

In the following, typical examples for energy losses will be presented, based on an analysis of one of the biggest companies. It has to be noted that the situation in many of the smaller, less-organized laundries is often worse. Besides the lack of knowledge, for many owners it seems to be simply not attractive to invest in measures to improve energy efficiency, since in absence of control by authorities, firewood is quite cheap.

Figure 3.6: Old (a) and newer (b) boiler systems to generate hot water and steam
(Source: own)

One point is the efficiency of steam boilers (Figure 3.6 on the facing page). Especially in smaller laundries, old boilers with a combustion efficiency of 60% or even below are still in use. More advanced steam boilers typically have combustion efficiencies of 70-80%.[4] According to Baudach (2007), the main problem is related to the excess of air during the combustion process. Another issue is the overcapacity of many boilers used, contributing to lower efficiency as well. As shown in the pictures, both types of above mentioned boiler systems consist of a furnace and the actual boiler part, while for latest-developed devices available on market both functional parts are combined, reducing additional heat losses. Therefore, replacement of the older boiler by modells of the newest generation with an adequate capacity might improve energy efficiency considerably, even though associated costs will be significant.

Furthermore, in many laundries steam tubes from boilers to washing machines and driers are either incompletely or not at all properly insulated. This leads to significant and unnecessary losses of energy. Even if tubes are insulated, then valves and connections are often not. Especially in smaller and less-organized companies, tubes are unnecessarily long and not insulated at all. The extreme case is a laundry in Caruaru, where the pipes cross a small channel and the street, given the long distance between the boiler and center of consumption. However, even though a proper insulation is neither complicated nor requires high investment costs, it contributes significantly to improve efficiency, with positive outcomes for the entrepreneurs as well as for the environment. Consequently, a proper insulation of heat pipes should be the first measure to apply, before taking into consideration any other. Due to insufficient awareness and insufficient cost controlling by the owners, this point often receives no attention at all.

Additionally, Baudach (2007) mentions potentials to reduce the amount of steam and electric energy used for the driers by avoiding too long drying processes. The average drying time of 70-80 minutes could be limited to the half. Installing control technology with sensors for temperature and humidity allow to stop the drying process automatically once threshold values have been reached. Further, applying heat changers allow to use heat energy contained in exit air in order to preheat air entering the machines.

Given the high water consumption, more advanced strategies to increase energy efficiency have to consider measures to reduce the amount of fuel used to produce steam. An anal-

---

[4] Information on combustion efficiencies according to Jefferson Silva, engineer of ITEP

ysis of the distribution of needed water temperatures (Figure 3.7) reveals that 54% of the total water consumption[5] refers to "cold" water, i.e. water of ambient temperature. Further 31% shares water with temperatures in the range of 40 to 60°C. Under these conditions, two very promising options are reusing the heat of of the water leaving the washing machines on the one hand, and solar energy for preheating fresh water entering the machines on the other.

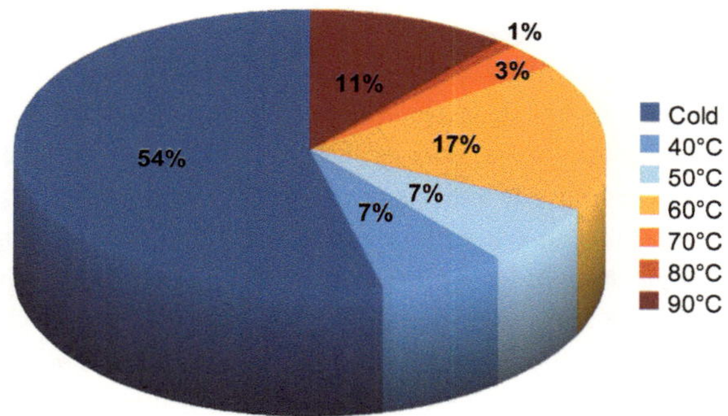

Figure 3.7: Distribution of water temperatures needed in a selected laundry
(Source: own creation based on personal communication with a laundry)

Regarding the first option, Baudach (2007) refers to a theoretical potential to reduce energy consumption of up to 40% by installing heat changers. Main obstacle lies in the complexity of applied washing processes, which complicates the separation of the used water. Another low-cost possibility could be using heat energy of the gas leaving through the chimney of the furnaces in order to preheat water before entering the boiler. In any case, the area of reusing heat energy deserves more research. The second option, using solar collectors, might be even more promising and also a visible sign of a more *ecological correct laundry*, at least in terms of energy issues. However, solar energy requires high investment costs, given the high demand of hot water. According to experts from the federal university UFPE, a 100% generation of hot water (up to 60°C) by solar collectors

---

[5] Referring to data delivered by the laundry, total consumption is in average 70 liters per pair of jeans and about 7 million liters each month.

for a big laundry needs an initial investment of approximately 400,000 R$ (Costa, 2007). Bearing in mind only the average annual fuel costs for wood (74,880R$),[6] the minimum pay off time corresponds to five years and four months.[7] Consequently, the vast majority of laundries is unable to apply this technology, even though fuel prices have to be expected to rise. A pay off after five years or even more is simply to long for many laundries due to unsure future projections.

Nonetheless, one of the biggest laundries shows high interest in solar power and currently plans to generate up to 50% of its demand on hot water by solar collectors. Main objective of the owner is to present a *ecological correct* way of laundering jeans in order to become certified after important worldwide accepted quality (ISO 9000) and environmental (ISO 14000) management norms (Costa, 2007). With regard to the company's role in promoting low-cost wastewater treatment technology in the region, this company might be again motivating other laundries to develop similar concepts. Furthermore, these improvements in the field of energy efficiency might be a chance for all stakeholders to exit this second *devil's cycle* regarding the unsustainable and often illegal cut firewood and offer a win-win situation for the benefit of the whole region. Textiles of high quality, produced ecologically correctly, may have a higher market potential and allow this for the region so important industry to sustain, without harming the environment.

---

[6] Based on monthly fuel costs by the use of thick wood of 6,240R$ (Figure 3.5 on page 17)

[7] This statical approach serves for illustration only, as it neglects maintenance costs for solar collectors and boilers, neither it includes interests of bank credits (which are quite high in Brazil), nor rising fuel prices that might favor the transition to solar energy. Labor costs are in the region are among the lowest of Brazil, and broken boilers might be cheaply repaired or replaced by second hand ones, that are available in abundance.

4 Future of Jeans Laundering in Caruaru and Toritama

So far, several first steps to minimize the environmental impact have been developed. As shown in the last chapter, this counts in particular for the treatment of wastewater and solid waste residuals. But there is still need for further improvements, mainly in the field of energy efficiency. To address these issues properly, the optimal solution might be the establishment of an industrial area for jeans laundries outside the centers of Caruaru and Toritama. This idea has been already under discussion for several years and is lately receiving more attention again. Main ideas include a central provision with steam, hot water and the treatment of produced wastewater.

Such an approach offers several advantages. Due to scale effects, solutions can be more cost-efficient and thus allow the implementation of better technology in the wastewater treatment station, as well as using renewable energies for heating up water and steam. Competition between the involved companies could not excuse anymore to not fulfill environmental standards, since steam and wastewater are provided central with same costs for all, but rather set further incentives to invest in more efficient devices. Clustering these activities in an industrial area would also allow small enterprises to work together and may join in cooperatives.

Additionally, efforts of involved authorities would be reduced, because they do not need to inspect dozens of laundries, each having separate boilers and wastewater treatment systems. In case the centralized services violated environmental regulations, e.g. by using illegal cut wood or improper treatment of wastewater, this would also draw much more attention of the public, and therefore all stakeholders involved have interest to avoid such scandals. Last but not least, if laundries move away from cities and neighborhoods, the impact on local population will be reduced, such as smoke and dust from furnaces, odors caused by wastewater etc. A scenery such as depicted in Fig. 4.1 on page 25, with "jeans-blue" water in a channel and a pig inside could not appear anymore, once laundries would leave the neighborhoods.

Unfortunately, there are also several problems connected with the establishment of an industrial area. Doubts on general feasibility are also shared among experts of ITEP. Most important issues seem to be financial aspects. The enterprises alone are rather unable to take responsibility for initial investment costs. Consequently, this leads to the question whether smaller laundries shall be allowed to remain in the towns, while only medium-sized and large-scale companies move to the industrial area, with all negative consequences this may have for the local population. The bigger the laundries, the more pollution they cause. But normally the size of enterprises correlates with possibilities to invest in treatment technology.

One option to obtain financial support could be to submit a project under an emission trading regime, such as the Clean Development Mechanism (CDM) established by the Kyoto Protocol to the United Nations Framework Convention on Climate Change. Main idea of the CDM is that entities from countries with obligations to reduce greenhouse gas emissions have the opportunity to invest in projects in developing countries, that have no reduction obligations. Resulting emission reductions can be traded and used by those that have to reduce their emissions. In such a way, a CDM project could for example contribute financing the use of renewable energy resources and measures to increase energy efficiency in the jeans laundries.

Another concern is the availability of data in order to provide sufficient capacity for steam generation and wastewater treatment. Again, this counts especially for small-sized laundries, that often have no exact information on water and steam consumption. Therefore, it seems to be sometimes surprising how they can maintain their activities. Furthermore, precise data are needed to develop a reasonable methodology for any climate change mitigation project under the CDM.

Nonetheless whether an industrial area will be built, the example of one of the bigger laundries (Chapter 3.3) suggests that there are a variety of possibilities to improve especially energy efficiency. This could assist not only to reduce environmental problems, but also to improve product quality. Therefore, most of the improvements could be also realized in the laundries without moving to an industrial area. So far, the vast majority of produced jeans is sold within the region. As textile imports from Asian countries are expected to increase, part of a successful strategy to continue with the local textile industry is to conquer other markets, in Brazil as well as in other countries. Producing textiles of higher quality improves chances to export. While lately entrepreneurs trying to establish the export to several African countries, *ecological correct produced jeans* of high quality may open opportunities to export to Europe or North America as well. This however also implies that consumers are willing to accept slightly higher prices.

Figure 4.1: Impact on the local population
(Source: own)

5 Summary and Conclusion

Brazil is one of the most unequal countries with regard to income distribution and land possession. As consequence, fast economic growth is often given higher importance than a more sustainable development. The the textile cluster around Caruaru in Northeast Brazil serves as telling example. For years, a *devil's deal* between authorities and entrepreneurs fortified the situation and inhibited any improvements of working conditions and environmental impacts. During the last years however, the situation started to change when businesses associations and authorities started to become more active to support the clothing industry including first steps to reduce severe environmental impacts of jeans laundering.

While aspects of high water consumption and pollution have already received special attention, the field of energy efficiency in jeans laundries has been left aside. There exists a huge potential to reduce the consumption of firewood by several measures such as reclamation of heat energy and solar heating systems, but an application of these technologies requires a relatively high initial investment. As long as illegal cut wood is available for a low price, entrepreneurs will be attracted to continue illegal practices. Still, also the example of wastewater treatment reveals a lot of potential for further improvement. One possible solution to tackle the problems of the jeans industry might be the establishment of an industrial area outside Caruaru and Toritama. Services like steam, hot water and wastewater treatment could be provided centrally for all involved enterprises. Many of the options to increase energy efficiency for all jeans laundering companies could be included. Yet, it is not completely sure whether this industrial area will be built.

On the other hand, some companies have already switched to alternative fuels and even show interest in applying solar heaters. Motivation behind is to become certified by environmental and quality management systems, such as ISO 9000 and ISO 14000, in order to promote environmentally correct produced jeans and be able to demand higher prices.

Given the pressure on the market by cheap imports from Asia, their promising strategy is increase product quality, so that jeans could be exported not only to other parts of Brazil, but also worldwide. In such a way, the whole region could benefit. Textiles of high quality, produced ecologically correctly, may have a higher market potential and allow this for the region so important industry to sustain, without harming the environment.

Concluding, these examples demonstrate that companies like jeans laundries can be convinced to invest in cleaner production most effectively by the help of economic arguments, when a clear link between economic benefits and environmental improvements can be presented. Difficulties arise from the interest in short-term profits, even though the higher costs related to environmental degradation in the long run. Therefore, I think this case study stresses the importance of components to provide environmental education and strengthen environmental awareness in the field of development assistance, rather than only present some technical solution for the problems. Finally, the aspect of environmental awareness also leads back to the responsibilities of consumers. Given the clear link between consumption behavior and environmental degradation, which often rather affects the poor more than the rich, this counts in particular for the consumption patters in industrialized countries.

References

Almeida, M. (2005). Understanding Incentives for Clustered Firms in Brazil to Control Pollution: The Case of Toritama. IPEA-Discussion Paper. Brasília, Brazil

Baudach, K.-M. (2007). Análise de fluxos de energia na lavandaria Mamute, Toritama. (Unpublished)

Cabral, S. (2007). Logística reversa estimula reaproveitamento de materiais. In Journal da Industria, Number 47, FIEPE, Recife, Brazil

Chapagain, A.K.; Hoekstra, A.Y.; Savenije, H.H.G.; Gautam, R. (2006). The water footprint of cotton consumption: An assessment of the impact of worldwide consumption of cotton products on the water resources in the cotton producing countries. In Ecological Economics, Volume 60, Issue 1

Costa, E. (2007). Lavandaria ecologicamente correta. In Diário de Pernambuco, Recife, Brazil (September 24, 2007)

IBGE (2007). Contagem da População 2007. (as of April 22, 2008)
URL: http://www.ibge.gov.br/home/estatistica/populacao/contagem2007/PE.pdf

Meier, R. and Wahl, M. (2005). How a business membership organization can contribute to improving the business environment. Paper presented to the International Conference on Reforming the Business Environment, Cairo, Egypt

Ramakrishna, K.; Jacobsen, L.; Thomas, R.; Woglorn, E.; Zubkova, G. (2003). Action Versus Words: Implementation of the UNFCCC by Select Developing Countries. The Woods Hole Research Center, Woods Hole, USA

Santos, E. O. (2006). Caraterização, biodegradabilidade e tratabilidade do efluente de uma lavanderia industrial. Master thesis at UFPE, Recife, Brazil

Seroa da Motta, R. (2002). Social and economic aspects of CDM options in Brazil. In International Journal of Global Environmental Issues, Volume 2, Numbers 3-4

Tendler, J. (2002). Small Firms, the Informal Sector, and the Devil's Deal. IDS Bullentin, Volume 33, No. 3, July 2002

UNDP (2003). Atlas do Desenvolvimento Humano do Brasil. (as of April 16, 2008)
URL: http://www.pnud.org.br/atlas/dl/unico/AtlasIDH2000.exe

UNDP (2007). Human Development Report 2007/2008. United Nations Development Programme, New York, USA